GUIDES

INFORMATION &
COMMUNICATIONS
TECHNOLOGY

REAL LIFE GUIDES

Practical guides for practical people

In this series of career guides from Trotman, we look in detail at what it takes to train for, get into and be successful at a wide spectrum of practical careers. The *Real Life Guides* aim to inform and inspire young people and adults alike by providing comprehensive yet hard-hitting and often blunt information about what it takes to succeed in these careers.

Other titles in this series are:

Real Life Guides: The Armed Forces
Real Life Guides: The Beauty Industry, 2nd edition
Real Life Guides: Care
Real Life Guides: Carpentry & Cabinet-Making, 2nd edition
Real Life Guides: Catering, 2nd edition
Real Life Guides: Construction, 2nd edition
Real Life Guides: Distribution & Logistics
Real Life Guides: Electrician, 2nd edition
Real Life Guides: Engineering Technician
Real Life Guides: The Fire Service
Real Life Guides: Hairdressing, 2nd edition
Real Life Guides: Information & Communications Technology
Real Life Guides: The Motor Industry, 2nd edition
Real Life Guides: Passenger Transport
Real Life Guides: Plumbing, 2nd edition
Real Life Guides: The Police Force
Real Life Guides: Retail, 2nd edition
Real Life Guides: Transport
Real Life Guides: Travel & Tourism
Real Life Guides: Working Outdoors, 2nd edition
Real Life Guides: Working with Animals & Wildlife, 2nd edition
Real Life Guides: Working with Young People

trotman

Real Life

GUIDES

INFORMATION & COMMUNICATIONS TECHNOLOGY

Colin Taylor

Real Life Guides: Information & Communications Technology

This first edition published in 2008 by Trotman Publishing, a division of Crimson Publishing Ltd., Westminster House, Kew Road, Richmond, Surrey TW9 2ND

© Trotman Publishing 2008

Author Colin Taylor

British Library Cataloguing in Publication Data
A catalogue record for this book is available from the British Library

ISBN 978 1 84455 170 5

Typeset by RefineCatch Ltd, Bungay, Suffolk
Printed and bound in Great Britain by Athenaeum Press, Gateshead.

Real
Life

GUIDES

CONTENTS

About the author

Colin Taylor is a Careers Adviser with the University of Cumbria who has spent many years working with young people and himself studied a number of degree-level modules in Computing. Since becoming a freelance writer in 2003, he has published a number of articles on careers and employment for the Association of Graduate Careers Advisory Services, jobs4U, *Nursing Standard*, *Nursing Times*, *Eurograduate* and Hobson's GET Guides. This is his first book.

He is also a qualified cricket coach who enjoys cycling, walking, angling and bird watching. When not outdoors or in front of a computer, he likes old comedy programmes, blues/rock and electronic house.

Acknowledgements

Thanks are due to all those who provided the author with information during the writing of this book.

Foreword

Information and Communications Technology (ICT) is part of everyday life, from your mobile phone to the computer you use at home. As technology advances our capabilities may not meet the requirements of employers and e-skills (sector skills council for IT & telecoms) see this becoming more of an issue over the next 5 to 10 years.

As IT becomes more pervasive in work and society our ability to use desktop technologies and understand information security requires us to have new and improved skills, which need updating regularly. The office isn't the only place that these skills are in high demand; combine harvesters, electrical testing equipment, even the tills we see in our supermarkets rely heavily on IT.

City & Guilds provides a number of ways to gain useful qualifications that have been developed with employment in mind, whether you are just starting out or looking for your next career move.

Does training work? One training provider found candidates who had achieved an iTQ (the NVQ/Apprenticeship for IT users) were reporting their new skills saved them on average 32 minutes a day. Taken over a full year this adds up to a productivity gain of 8% – that's 120 hours over a full year or 3 weeks extra work.

City & Guilds is delighted to be part of the Trotman *Real Life Guides* series to help raise your awareness of these vocational qualifications. If ICT is for you, all you need is the right training, so for more information about the courses City & Guilds offer check out www.cityandguilds.com – get yourself qualified and see what you could do.

Ken Gaines
IT User Product Manager
City & Guilds

Introduction

It is often said that, with today's fast pace of change, a job no longer lasts a lifetime. Well, the Information and Communications Technology (ICT) industry is so young that most of the careers we associate with it have been in existence for much less than a lifetime. Fifty years ago, there were no such careers as web developer or web designer as there was no internet. There were no games developers as there were no games consoles and the computers that existed were too unsophisticated to need network or database managers.

HOW IT ALL BEGAN

The first modern computer was built just after the Second World War in Pennsylvania, USA and was called ENIAC. ENIAC had to be programmed by six women who manipulated its switches and cables and it had no memory. By 1952, IBM had produced the first mainframe computer in its 700 series. Shortly afterwards FORTRAN became the first computer language and the age of the big mainframe, shining in its huge air-conditioned laboratory, slowly came into being.

By the late 1960s, the first computer job titles started to emerge, such as computer engineer, systems analyst, computer

DID YOU KNOW?

As early as 1834 Charles Babbage invented, but never completed the analytical engine, a forerunner of the modern computer. It performed arithmetic calculations using punched cards to provide instructions and had its own memory unit.

programmer, and computer operator. These jobs still exist today, although you will more likely hear a programmer referred to as a software developer (or similar) and the term business analyst is becoming more common than systems analyst. Other jobs from this era, like punch card operator or tape librarian, had a very short shelf life and no longer exist.

THE GOOD OLD DAYS

'When I first started as a programmer in the early seventies, I was lucky enough to be working on programmes that were input on punched cards. Most people were still using paper tape. If there was a mistake, they had to reel back the tape to find where it was and splice in a new section with the corrections.

'When the first IBM PCs were introduced, they cost £4,000 each and, we were amazed that some of them had as much as 40 megabytes of disk storage space. We used them for word processing, spreadsheets and project planning. At that time, they weren't used for programming at all.'

Alan Gray, now Project Manager for EDS, Sheffield

THE INTERNET AND THE WORLD WIDE WEB

The first IBM PC became available in 1982, by which time simple games consoles and machines like the Sinclair ZX, which could accept ordinary programmes but took ages to

> **JARGON**
> **ENIAC** – Electronic Numerical Integrator And Computer.
> **FORTRAN** – IBM Mathematical Formula Translating System (an early computer language mainly used by scientists and engineers).
> **ARPANET** – Allied Research Projects Agency Network.
> **NSFNet** – National Science Foundation Network (USA), an offshoot of ARPANET.
> **gif** – graphical interchange format (a good way of storing illustrations, but not so good for photos).
> **Wiki** – what i know is (and a way of showing what you know on the web!).

process them, started to appear in people's homes. Through the nineties processing speeds improved resulting in faster, more efficient computers, but it was the internet that made information technology a part of almost everyone's life.

The first *internetwork protocol* came from the USA Defence Allied Research Projects Agency in 1969 (later known as ARPANET) so researchers could share military information. It linked just four mainframe computers. By the 1980s NSFNet provided a separate network of computers used by universities and scientists. Some of them were not very pleased when, around 1984, NSFNet was opened up to other users and businesses started charging users to access it, but it was NSFNet that started the use of the internet for fun and information as well as business. From around 1990, growth shot through the roof and there were 34,600,000 users of the internet by 1996.

Growth of the internet in 12 year periods

1972	ARPANET	37 computers
1984	NSFNet	5,000 computers
1996	Internet	34,600,000 users

The main reason for the explosion was the invention of the World Wide Web by Briton Tim Berners-Lee in 1989. Up to this point, you had to access all internet information from a UNIX command line – a very unfriendly process for everyday users.

Often confused with the internet itself, the World Wide Web (www) is actually a piece of software that lets you view pages through a page window that shows hyperlinks.

As the www became more sophisticated, the internet landscape of today started to take shape, with internet service providers, browsers, search engines, e-commerce applications and e-learning systems – as well as all the new IT jobs that came with them.

THE DOTCOM CRASH AND WEB 2.0

The period from 1970 saw a massive growth in computer jobs that was only halted by the catastrophic dotcom crash of 2000. The crash led to lots of bankruptcies and redundancies in ICT employment after too many people invested heavily in new internet companies.

DID YOU KNOW?

In 1971, Ray Tomlinson, a computer engineer, wrote the first email message when he sent a series of test messages to himself from one machine to another. He is quoted as saying: 'Most likely the first message was QWERTYIOP or something similar.'

Source: www.hodgy.net

SEARCH ENGINES OVER THE YEARS

In the early 1990s the most popular search engines, such as Webcrawler, were basically directories, so that you had to search the file structure to find what you wanted.

Google became the most successful search engine of the new millennium through the introduction of keyword searching, allowing it to search millions of sites in seconds.

Web 2.0 engines like delicious ask you to bookmark the sites you like and tell you about sites other people like. StumbleUpon brings up a random series of sites that it thinks may fit your stated interests.

Britain was late to be affected when, according to e-skills UK, jobs in the ICT industry sector fell from a peak of 1,033,000 in 2001 to just 955,000 as at February 2002. 25,000 ICT workers lost their jobs in the period from September to November 2001 alone.

Recruitment then increased again over the next five years, but the dotcom crash does serve as a warning that jobs can appear or disappear equally quickly in a fast moving industry such as ICT.

Web guru Dale Dougherty and Tim O'Reilly of oreilly.com first used the term *Web 2.0* at a conference. Since 2005, it has been used to describe a new breed of websites that share two key characteristics:

- Web 2.0 sites use an attractive variety of formats, including complex graphics, video and audio, rather than just text. A dozen years ago websites relied mainly on text and simple gif images.

- High levels of user involvement. Whereas early websites were like online books, written by a design team and read by the user, Web 2.0 sites invite you to contribute to them by commenting on them and providing your own content.

As well as well-known successes of social networking sites such as YouTube, Facebook and countless blogs, we now have search engines and even anti-virus products that adapt themselves according to the needs of users, wikis helping people with no technical knowledge to write webpages, and sites where you can store all your documents and invite people to comment on them.

Although cynics see the seeds of a new dotcom crash in Web 2.0, there is no doubt that it has given the internet and the IT job market a new push, and has influenced the skills needed by companies.

THE INDUSTRY TODAY

Today's ICT industry is even more high pace and fast changing than it ever was, and has become a massive employer. For many, it is an exciting place to be.

According to e-skills UK, over a million people now work in information technology (IT – ICT minus the telecommunications sector) but less than half work directly for IT companies, with 55% working as specialists in companies that use IT. Of the whole UK workforce, 5% now work in jobs in ICT and 77% of all workers frequently use computers in their job.

Although most computer equipment is no longer made in the UK, very large numbers of people work in areas such as management, planning and strategy. Software development and internet services offer lots of opportunities and there are still lots of people working in areas such as IT sales, user support, database administration and engineering. The UK is also known for its thriving computer games industry.

For more information on trends in ICT, see Chapter 2 'What's the story?'

DID YOU KNOW?

CompuServe was already offering electronic mail and real-time chat to personal computer users by 1980.

Source – Wikipedia

WHAT'S IN THIS BOOK?

In the rest of this book you can get a breakdown of some of the main jobs offered by the industry today. You will also find out about the skills and experience you will need to succeed, what the conditions of work are like and an idea of what makes computer people 'tick'. The case studies throughout the book add a personal touch, dealing with real experiences of real people. The different sections should help you decide whether you are cut out for a career in ICT. If in doubt there are some simple checklists and quizzes you can use to see how you measure up.

If you are starting to think that ICT could be for you, the book goes on to tell you about applying for computer jobs, including hints on how to complete online applications and find information on writing a CV – both very important for ICT jobs. It also talks about the training opportunities and courses available, especially IT Apprenticeships and the new IT Diploma now available in schools.

If you are fascinated by computers and how they work, maybe you too will find your way into the ICT industry, perhaps even in a job that has yet to be invented!

1

Success story 1

STUART IS A FAMOUS EMPLOYEE

A career in ICT means you can be a big success in a relatively short period of time. Stuart Blenkhorn was only 18 months into his current job as a Senior Desktop Support Technician with Skipton Building Society when he was selected as the company's 'Famous Employee of the Year' for 2007, and he's still only 26 years old.

Stuart, who is from Cononley, North Yorkshire, started off on a media studies course at Craven College after leaving school before he decided to try for ICT jobs. He says:

'After the course I was selling pork pies for the family firm and went back to college to do another course for the European Computer Driving Licence. I'd always been interested in computing in my spare time and someone on the course tipped me off that there was a job coming up at Skipton Building Society. I got the job doing first line phone support and moved on to desktop support before getting promoted to the job I do now.

Stuart is extremely customer focused, regularly putting himself out for others.

'Support technicians deal with the most difficult cases as a backup to the telephone helpdesk. We go out to see the customer to deal with whatever technical problem they have – it could be anything from computers to multipurpose printers or Blackberry handsets. As a senior I also have to show new staff the ropes and deputise for my team leader.'

So how did a support technician win the sort of award that usually goes to long-serving company managers or high-profile marketing staff? Stuart explains:

'First I think awards are a really good idea. What happens in our company is that, every three months, we all vote for the person who we think has done most to go beyond the call of duty in their job – in each of three separate categories. One quarter, I came first in the "Customers love us" category and won a nice stay with my wife in a hotel in Edinburgh. As a result I got invited to our annual conference in Chester, which is when the annual award is announced.

'The only trouble was that I went down with a nasty stomach bug just before the conference and couldn't go! Then lots of people started ringing me up and sending messages to say how important it was for everybody to attend the conference. But I just assumed they thought I was trying to get out of it. I actually had no idea I'd won "Famous Employee of the Year" until I got a text from Steve Aldous, our General Manager. Apparently he'd given his mobile to his secretary before going on stage to announce the awards and asked her to send the text the instant mine was announced so I wouldn't find out from anybody else!'

This time Stuart won a short break in Dubai – but has decided he'd rather take his family to Euro Disney instead.

'It's a pity I couldn't receive my award in person. I wouldn't have minded giving a speech and I love watching the conference video where I receive a standing ovation in my absence! But it is really satisfying to think how much harder it is for a technical person to win an award like this. It puts our support team on the map and shows how important it is to work well with internal as well as external customers. Most of my votes were in the "Customers love us" category.'

FROM STUART'S CITATION FOR THE AWARD BY STEVE ALDOUS

'Stuart is the choice of myself and members of the Operational Board, for 2007 "Famous Employee of the Year", for his consistent, enthusiastic and reliable behaviour. Very much one of the troops, Stuart works as part of the IT desktop support team and has a strong reputation around the business for the excellent level of service he provides to his customers – other members of staff.

'During the conference on Saturday a lot of emphasis was placed on the importance of every one of us delivering first class service at all times and I can't think of a better example of this than Stuart. He is extremely customer focused, regularly putting himself out for others. Nothing is ever too much trouble and he always takes responsibility and ownership in attempting to solve any difficulties his customers are experiencing.'

JARGON

ECDL – European Computer Driving licence, a qualification in basic computing skills you can do to improve your understanding of ICT.

External customer – Someone who buys a product or service from a company.

Internal customer – A colleague within the company who helps ensure that external customers receive high quality products and service.

2

What's the story?

We've already said that over a million people work in IT jobs, and if we include the closely related field of telecommunications, the number actually goes up to 1.5 million ICT employees. We also mentioned some of the main broad job areas people work in.

Now let's look at some of the facts and trends in the industry, so you can see a bit about what is going on behind the scenes. A lot of this information comes from e-skills UK, who are the sector skills council for the industry, and you can read their detailed reports at www.e-skills.com.

A GROWING INDUSTRY

Forecasts made in 2007 predict there will be 2.5% more IT jobs by 2016, a bigger rise than for UK jobs generally, so it is good news for would-be IT buffs. If you look at the figures, though, it becomes clear that some of the areas of employment we talked about are growing faster than others. The not-so-good news is that the main growth is in IT management, strategy and planning and software development, which all tend to need high levels of skill and technical experience. Jobs in other areas, like computer engineers and database administration, will remain at roughly the same levels, or even fall slightly.

The following diagram gives a rough picture of how the main sectors may fare over the next ten years.

Going up, going down – IT employment sectors 2007–2016

Large rise in numbers	Annual change	Small rise in numbers	Annual change	Small fall in numbers	Annual change
IT managers	+1.4%	Operations technicians	+0.8%	Database assistants and clerks	−0.8%
IT strategy and planning	+2.6%	User support	+0.7%	Computer engineers	−0.6%
Software development	+2.7%				

Source: e-skills – 'Technology Counts' report

Fortunately, there are no big IT job losses likely over the next ten years.

THE NUMBERS GAME

Now here are some key statistics about jobs in the industry today:

- You are much more likely to connect with a job in ICT if you apply for jobs in the London or the South East of England, as 42% of all ICT jobs are based there. The worst areas for ICT jobs are North East England and Wales, with only 2% of the ICT workforce each.
- Most jobs are in very large or very small companies. 50% of people work in just 328 large companies but 86% of companies employ four people or less.
- Many people are self-employed – 24%, which is much higher than the UK average. Consultants rule!
- There are a lot of graduates in the industry and 55% of all ICT employees are graduates. Of course this doesn't mean they all went to university as soon as they left school.

- At the moment, employers seem to prefer older applicants who have done other jobs and few really young people work in ICT. Only 1% of the workforce is aged under 18. So it's likely that you'll try something else before going into ICT, perhaps picking up computing skills along the way.
- Only 18% of ICT employees are women, far lower than the UK average of 47% and still falling. Employers are starting to get worried about this trend and there is a lot of help available to get girls more interested in ICT jobs.
- More employees (6%) describe themselves as 'Asian' or 'British Asian' than most other industries.

OPPORTUNITIES FOR YOUNG PEOPLE
Stephen Uden, Head of Skills and Economic Affairs, Microsoft UK, believes there will be plenty of opportunities for young people in ICT:

'The ICT industry will need around 150,000 new staff a year over the next few years and I hope that half of those will come direct from education. Not all opportunities will require high-level skills – for example Helpdesk jobs, the IT Apprenticeship and the new IT Diploma.

'I also expect 5,000 new businesses to start up, mainly operating online internet sites that young people understand better, using multimedia content, social networking and search tools.

'Skills shortages in the industry mean that you can move upwards very quickly regardless of how you got into it. One of our managers at Microsoft was promoted four times in four years.'

CRYSTAL BALL GAZING

If Gypsy Rose Lee looked into her crystal ball to work out exactly what is happening in the industry just now, it would probably turn quite murky and fill up with complex arguments and fancy buzzwords. Let's try and explain some of them:

Globalisation and offshoring

Because most countries have good internet access, companies can get a lot of work done more cheaply abroad nowadays. This partly explains why numbers in some jobs are falling (see above). The obvious example is using call centres and programmers/developers in India. It also means UK companies can take contracts from across the globe. Many ICT staff work in other countries too.

Convergence and standardisation

If you access the internet on your mobile phone, it's an example of *convergence* – the trend to make more and more facilities available on different devices. So you can also watch television or send texts from your PC, or make free phone calls using Skype. This means lots of people have been sorting out software problems to ensure that the devices can all use the same protocols, i.e. speak roughly the same language – *standardisation*.

These developments are bringing IT and Telecoms closer together. Key areas are NGN/VOIP and WIMAX (see 'Jargon' box on page 19).

Green IT

Unlike the fortune teller's crystal ball, computers use an awful lot of electricity, so hardware and telecoms specialists are always looking for ways to make them run economically to prevent global warming, and reduce any harmful radiation from devices that could damage people's health.

At the same time, the Government is fighting hard to reduce the amount of paperwork that gets printed by promoting 'electronic government' and e-signatures. All businesses are aiming for the 'paperless office' – but few have got very near achieving it yet!

TECHNOLOGY TASTERS

- Following on from intelligent fridges and ovens, the intelligent toilet has now been invented in Japan.
- The days of the phone bill could be over. Telecoms companies are ready to route phone calls over the internet instead of the phone system.
- Digital tagging (implanting microchips under the skin) may mean ID and credit cards will no longer be needed, as computers will scan the user not the card.
- Radio Frequency Identification (RFID) could mean that passers-by will soon be able to point a mobile phone at a poster or advertisement and download details of the product from a minute chip in the ad.

Source: prospects.ac.uk

Data protection and cybercrime

At the time of writing the papers are carrying stories about sensitive personal information being lost from computers, laptops or disks almost every month. Meanwhile, the very success of the internet has led to a vast increase of criminal activity involving computers. Identity theft, computer scams, phishing, abusive porn sites and illegal downloading are probably the worst examples.

No wonder that banks especially are spending millions of pounds on trying to combat computer fraud with new technical solutions and that 'forensic computing' has become a full-time job for teams of experts in the police force and elsewhere.

Social computing and digital natives

Most readers of this book will be digital natives – people born after 1985 who grew up with computers from an early age and make fluent use of social websites like YouTube and Facebook – both part of the Web 2.0 revolution mentioned in the Introduction.

Companies are putting a lot of effort into understanding how they can tap into this new group of potential customers – so maybe they do have some use for young people after all!

> **DID YOU KNOW?**
>
> A data centre with 25,000 servers will use enough electricity in a single month to power 420,000 homes for a year.
>   Source: National Computing Centre

Analytics

Computers and the internet are becoming more and more central to the way companies do business, so computer

---

KEY FACTS
- There will be 2.5% more IT jobs in 2016.
- 42% of all ICT workers work in the South East.
- 55% of people in the industry hold a degree.
- Only 18% of ICT staff are women.

---

experts need to know more about how businesses operate. Analytics is about gathering intelligence on the way a business works, and suggesting how technology might open up new opportunities to make money or improve service.

AND NOW A QUESTION FOR YOU

Did you enjoy finding out about some of the trends in the industry and what some of the 'buzzwords' mean? If so, then maybe you have the sort of enquiring mind that will enjoy finding out how to fix a bust network or develop a new piece of software.

One thing is certain. The ICT industry will continue to change at least as quickly as some football teams look to change their managers. Many of the trends people talk about as you read this book will take off and become huge successes, but some will end up on the scrap heap in double quick time – just like all those old dotcom companies we talked about. It's not a place to be if you don't like lots of change.

JARGON

WIMAX – worldwide interoperability for microwave access to let all devices use wireless connections as alternatives to cable and DSL lines. In particular, a way of getting very high speed broadband into buildings without replacing miles of telephone wire.

VOIP – Voice Over Internet Protocol. A way of transferring speech and audio over the internet. Maybe one day we will even be able to dictate messages into the PC rather than typing them!

NGN (or C21N) Next Generation (or 21st Century) Networks. An umbrella term for networks using a mixture of data and VOIP.

E-signature – A way of producing legally binding official documents electronically, without the need to sign printed copies but also without the risk of fraud. Really a form of e-commerce, the UK online tax return is an obvious example.

Phishing – Trying to get someone's account details by sending them a bogus email supposed to come from their bank or other organisation, usually saying they have to update their account details.

Forensic computing – The use of computer expertise to trace or prevent cybercrime.

3

ROB LANGER

Case study 1

MEDIA TECHNICIAN

Rob Langer is a Media Technician with Staffordshire County Council. He designs publicity and other publications for use in schools and helps with technical support on equipment and programmes.

He completed the Level 2 and Level 3 Apprenticeships in Information Technology, a programme lasting around two years.

How did you first get interested in an IT career?

'At school I got good GCSEs and stayed on in sixth form to gain a few average A levels. I never really planned to go to university as I wanted to get out into the world of work and earn a wage. Upon leaving sixth form I applied for the police but failed the eyesight test, which was a bit of a knock back as I had set my heart on it for a while.

'I then looked at a career in graphic design and that's when I discovered the IT Apprenticeship, which was part of my training when I joined the Media Department at the council.'

The good thing about an apprenticeship is that you get hands-on training in a working environment, which will give you vital experience.

What did you learn during the IT Apprenticeship?
'It covered a wide range of areas that includes using
Microsoft Office programmes like Excel and Publisher,
basic IT systems and peripherals, IT security, etc. My IT
tutor provided projects on various IT-related areas for
each part of my apprenticeship portfolio. They were easy
to understand and the workbooks were presented in a
step-by-step style.

'Alongside the apprenticeship I learned how to use graphics
software packages like Photoshop, QuarkXPress and Corel
Illustrator and some of my projects were about designing
publications. I would have been given different projects if I
had been working in another department.

'The good thing about an apprenticeship is that you get
hands-on training in a working environment, which will give
you vital experience. Even if you are not very computer
literate, the apprenticeship will increase your knowledge and
understanding.'

How easy was it to get an IT job?
'I wouldn't say it is easy to get a job but I will say that there
are lots of opportunities for someone who is IT qualified. It is
definitely the industry to be in and there will always be jobs
available.'

What are the main pros and cons of your job?
'I enjoy the creative satisfaction of producing publications
and hearing from satisfied customers. I also get to learn a lot
about what happens outside – like how printing firms work
and what they need from me. There is good teamwork here
in the department and we all back up each other's job roles,

which is how I got involved in technical support and I also take photos sometimes.

'To be honest there aren't many cons. Of course there are software problems that need sorting out, or a customer has requirements that the software can't handle, but I enjoy the challenge of finding an inventive way round the problems.'

What are your ambitions for the future?
'I would like to become an IT support officer and perhaps even a server support technician. The earning potential is great with a basic starting salary of £22,000. Plus there are quite a few jobs like these around.'

What advice would you give to young people considering an IT career?
'If you are interested in IT and leaving school and a little unsure of what to do, and you feel lost in a sea of college this and university that ... choose an IT Apprenticeship. Once you have completed it you can go into a more advanced IT area. Take on board what teachers say, but do not think that you have to go to university – it is not the end of the world if you don't. There are plenty of places that offer the apprenticeship – I would advise contacting local authorities, etc. They offer good ones.

'You can't always expect a massive salary but remember that you are training for greater things and getting paid, whereas at university you will pay a high course fee and there is no guarantee you will like it or be successful. At least with an IT Apprenticeship you don't really have anything to lose. Plus you have hands-on support from an experienced tutor and colleagues who will help you through.

'I would however advise people to consider going to college part-time after or during the apprenticeship. I am currently applying for a server support technician course that will boost my earnings and prospects. I would not have had the knowledge and understanding needed for this course without my IT Apprenticeship.'

A QUESTION FOR YOU
Would you have thought of getting into ICT by applying for design jobs? Rob's story shows how you can move into ICT gradually, because ICT is such a big part of most people's work.

4

What are the jobs?

Of course there are lots of them – but first a word of warning! If two experts sat down and wrote a list of all the jobs in the industry, they would not write exactly the same headings. Because ICT is quite new and very fast-moving, job titles are not standard across the industry. They often overlap. The same job can have different job titles and people with the same job title can be doing different things in different companies, so it can all get a bit confusing.

In this chapter, we look at some typical job titles and try to classify them in rough groups to make it a bit simpler. It's also important to remember that a lot of the jobs in the industry require previous experience and training. The ones that don't require extensive experience – the so-called 'entry-level jobs' – are highlighted *in bold italics*.

DATA PROCESSING
When you sit in front of your PC or laptop and start to type, you are producing data that gets recorded and maybe stored and processed. These days, most data is directly produced by customers (e.g. if people order something over the internet) or office and other workers as part of their daily job. But there are still jobs for *data entry clerks, VDU operators* and *data processors*, whose basic job is to input and retrieve data collected by other methods to and from computer systems.

Once data has been classified in a database, *database assistants* import and export selections of data (e.g. for reports) and **data analysts** use advanced techniques to analyse data trends that might help management make decisions. It is absolutely crucial that databases run smoothly, so there are **database managers** and **administrators** to make sure that the data they hold is available, accurate and secure. They also decide who can access what data and so need a good understanding of the uses of the database. **Database programmers** write programmes in specialised languages (e.g. SQL) that are used for setting up and searching databases.

DID YOU KNOW?

Web Designers need to learn that people read web pages differently from books. They tend to scan for key phrases, starting from the left and working first across then down.

Source: WebPro News

CUSTOMER AND USER SUPPORT

If your new PC crashes or a programme won't run and you need to speak to someone, it will be someone like a *helpdesk analyst* or *support technician*. Support roles are varied: they can cover hardware or software problems and these days probably a bit of both. They may involve visiting users or offering remote support by telephone or e-mail.

Most ICT jobs involve contact with other people but other jobs where people skills are at least as important as technical knowledge include **sales consultants** who may be selling complicated new ICT systems and *sales assistants* selling computer equipment in the local retail park.

ICT trainers use their own ICT knowledge to help others learn how to use particular applications, and this includes teaching college lecturers using virtual learning environments to teach students. **Teaching and lecturing** are themselves big areas of employment for those who hold sufficient qualifications.

HARDWARE AND SOFTWARE

You probably know that hardware is all the mechanical and electrical parts that make up a computer and its peripherals, whereas software consists of all the programmes that make the computer actually do things.

Computer engineers design, build, test and repair computer hardware such as chips, circuit boards, disk drives, printers. One of the more technical roles in the industry, it is really a form of electronic engineering and needs the same level of technical interest.

Although the PC now dominates, *computer operators* are still needed to load and monitor programmes running on large mainframe computers and backup systems at the end of the day.

On the software side, we have an army of **software engineers** and **developers/designers** (formerly **programmers**) using an immense number of programming languages so that users are able to do virtually anything on a computer except eat, sleep and travel. Logic and problem solving is the name of the game here, but not at the expense of poor people skills. Some of the key software languages at the time of writing were C, C#, Java and ORACLE – but it could be different by the time you start work!

CONFLICTING LANGUAGES AND SYSTEMS

C# (C sharp) and **Java** are competing languages that both aim to improve productivity of web applications.
ORACLE is the leading database language, but is rivalled by **SQL**.
LINUX is a challenger to **Windows** and thought by some to be more secure.
Web Browser **FIREFOX** was voted by readers of one magazine as better than the more familiar **Internet Explorer**.

Network managers install and maintain computer networks, where computers are joined together in a local (LAN) or wide area (WAN) network, linking PCs and laptops to the servers that house the network facilities. Security and backup are crucial and the manager must understand the software and hardware involved. This is not a place to be if you panic easily when things go wrong. Meanwhile, **systems administrators** have a detailed knowledge of the specific operating systems, such as Windows and Linux that are used to run computers and networks. They sort out detailed problems to make sure that operating systems are working properly, including firewalls, routing and IP addressing.

Business analysts analyse what a business does using modelling and mapping and suggest how ICT solutions can help the business improve. They agree project schedules and may co-ordinate the actual projects. You need to have a high level of understanding of both business and ICT and generally be a bit of a whizzkid. A similar job is **systems** or **technical architect**, but usually with more stress on implementation.

We said in 'What's the story?' that there are an increasing number of ICT management posts. A typical example is the **project manager**, whose job is to co-ordinate the different parts of large ICT projects, making sure that all parts of the team work together so that the customer gets what they want at the end of the contract.

THE INTERNET

The internet is now so dominant that all ICT professionals need to know something about web technology. *Web designers* deal with the content, layout and coding of web pages – the 'front end' that you actually see. **Web developers** programme what needs to happen at the 'back end', for example in dynamic and database-driven websites. Increasingly the two roles are combined. **Web authors/ editors** specialise in writing text articles for websites and **interactive media designers**, who can also work with offline media like television and DVDs, use graphics, sound and animation to produce the effects that turn websites into little cinemas.

All these roles are a mixture of the technical and the creative. On the marketing side, **SEO (search engine optimisation) consultants** aim to improve a site's search engine listings, while **PPC (pay per click) consultants** look to get the best from the growing 'pay per click' internet advertising business.

COMPUTER GAMES

Focussed on Scotland, the North West and the Midlands as well as London, the UK has a strong computer games industry. **Games designers** devise the basic concept of the game including the characters, setting and storyline. Specialised **programmers** create the code to run the game and *level editors** develop the set and story for one part of game, using specialist 3D packages to create the exciting challenges the hero has to face.

Quality assurance technicians or *games testers* are employed to test and debug prototype games and suggest minor improvements. It isn't just about playing the games as you have to look out for continuity and all kinds of other mistakes too.

There are also opportunities for creative staff like **artists** and **animators** as well as *assistant producers** who check the progress of different parts of the project.

MANUFACTURING AND TELECOMMUNICATIONS

Very little computer equipment is now made in the UK, but in some places there are still a few jobs for *electronics assemblers,* **electronics engineers** and similar engineering roles.

---

* A little previous experience would be usual for these posts.

The related telecommunications industry does have a lot of opportunities for ICT roles. *Telecommunications technicians* and **engineers** also need to understand a lot about ICT systems such as broadband, networking and firewalls, while *cable jointers* and *line repairers* work on telephone and power lines.

ARMED SERVICES

There are a number of technical and telecommunications trades available for those who are also committed to the armed services, including for example *information systems engineer* (army), *communications information specialist* (Navy) and *ICT specialist* (RAF).

SELF-EMPLOYMENT

We shouldn't forget the large number of people in the industry who are self-employed. This can cover anything from **business analysts** and others who decide to set up as **independent consultants** to **freelance** web designers, or people who set up as *internet traders*, providing a wide range of goods and services using their own website or through eBay and similar sites.

MORE INFORMATION

The mindmap diagram opposite shows you most of the jobs we have talked about at a glance.

If you like the sound of any of these jobs, why not find out about them in more detail using the 'Further Information' section. Remember that, even if you don't end up working in ICT, 77% of all jobs require the use of a computer so your knowledge of computers will always be useful.

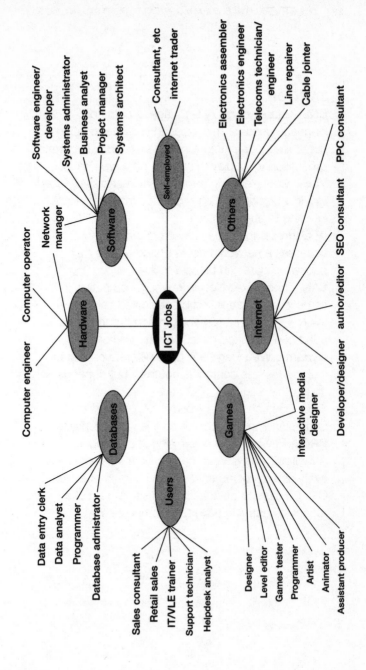

JARGON

SQL – standard query language, a common language used for manipulating and interrogating databases, often in conjunction with **PHP** (**h**ypertext **p**rocessor) to display the results on a web page.

VLE – virtual learning environment, increasingly used to allow students to learn over the internet, without needing to attend a class.

IP addressing – Every piece of hardware that accesses a network has an IP address so that packaged data can be sent to it.

LAN – Local Area Network, where computers in the same location are linked by dedicated cabling.

WAN – Wide Area Network, in which computers far apart can be linked by telecommunications.

Dynamic Web Page – A page where, unlike static web pages, the content is not fixed but depends on another source.

SEO – Search Engine Optimisation, the art of understanding how search engines rate websites.

PPC – Pay-per-click advertising (e.g. Google Adwords), where a company pays a small amount every time a customer clicks on an internet advert. Can be used to drive traffic to a site or to make money by carrying adverts for other sites.

5

LOUISE MITCHELL

Case study 2

ICT TRAINING PROGRAMME MANAGER

Louise Mitchell is a Programme Manager with Zenos, an ICT training company based in Banbury. Zenos deliver Advanced Apprenticeships for IT professionals in the ICT support industry as well as the new Train to Gain ICT programme for adults, ICT vendor qualifications (see 'Training for Tomorrow') and some commercial training.

Now 30 years old, Louise has been managing the Zenos apprenticeship programme for over a year. She is responsible for the delivery of the programmes for work-based learners and visits companies anywhere between Birmingham and Newcastle, and also works in the Zenos Academy, a chain of 12 training centres providing off-the-job training for the young apprentices.

How did you get into ICT?
'When I left school I didn't know what I wanted to do and ended up as an office junior on an apprenticeship in administration at Ingersoll Rand in Bolton. I quickly found out that I have a very good memory. When the ICT support staff

My degree covered all kinds of possibilities like system development, programming and database management, but I loved dealing with people, enjoyed training and wanted to do more.

visited our office, I always wanted to find out what they were doing so I could do it myself. Soon all the other people on the customer service team I worked for started to ask me how to do things on their computers too.

'Then the company changed over from dumb terminals and there was one gentleman who was very worried about using the new PCs. I spent a lot of time helping him to use his new system and realised I was good at showing people how to do things.

'By the time I completed my Level 3 in Admin I was looking to do an HND in Computing but found out I could get straight onto a degree at the University of Bolton. Even though I wasn't doing an ICT job at the time, Ingersoll Rand agreed to cover my course fees and for the next four years I spent two nights a week attending lectures after work for my degree in computer science.

'After about a year of the course I got a job in the Computer Operations Department at Ingersoll Rand doing helpdesk and backup support. In time I got more involved in support for UNIX and the MFG Pro front end, so I was doing some system administration tasks like running scripts for overnight batch processing, system backups, working with the development team to implement system changes and controlling access permissions. I also maintained and updated equipment and did some systems analysis when we migrated to MFG Pro.'

How did you get involved with training?
'I enjoyed my time in IT Support as I had always been looking to get more involved in ICT. The only thing I didn't

like was the physical side – carrying monitors around and crawling under people's desks. My degree covered all kinds of possibilities like system development, programming and database management, but I loved dealing with people, enjoyed training and wanted to do more.

'So after five years in ICT with Ingersoll Rand, I moved to Zenos as a Programme Facilitator. The job involves mentoring trainees – working one to one with young apprentices in different companies as well as observing their performance, assessing their progress and giving them feedback. It includes coaching them in "soft" skills like dealing with people and business awareness as well as the technical side.

'I still mentor trainees as Programme Manager, but now I'm also responsible for managing staff, monitoring the quality of their work, the training materials and programme delivery overall.'

What's good and what's not so good about your current job?

'Firstly it's extremely rewarding to have come full circle and see young people succeed in an apprenticeship after starting on one myself. I believe the Zenos programme provides young people with a fantastic opportunity and if it weren't for the apprenticeship programme, I wouldn't be where I am today.

'Although I like driving, I do cover a very large geographical area working from home in Bolton. A 9 a.m. meeting often means I need to start out at 6 a.m. and I have little spare time to myself.

'Some of the apprentices in call centres get a bit demotivated if they are just logging complaints all the time but I encourage them to think of their long-term development and the positive impression they can make on employers if they continue with the programme, develop their skills and gain the qualifications it offers.'

How do you see your future?
'At the moment I'm thinking of doing a Master's degree in Education in IT if I can spare the time.

'Overall I see myself getting more involved in the management of training programmes, writing new training materials and developing them for the marketplace.'

Any advice for young people thinking about going into ICT?
'At interview employers will look at how good you are at dealing with people as much as your technical skills. You also need to come over as positive, focussed, hard working and interested in developing your skills.'

JARGON

Dumb Terminal – A terminal that has no processing power and can only display information processed by its host computer.

UNIX – Operating system used to control computer networks linked to a server.

MFG Pro – An integrated frontend system used by companies that make and supply goods.

6

Tools of the trade

Now that we have had a look at the exciting range of jobs available within ICT, it's time to look at the skills you will need to make a successful career in this fast-moving industry.

If you are keen on a particular job, you can get a detailed idea of what's required by looking at careers databases like jobs4U, the online Careers Advice Service, Careers Scotland or Careers Wales. In this chapter we will look at skills needed throughout the industry or for different kinds of jobs, with a few examples along the way.

First think of the skills needed for ICT in four different main areas:

Tools of the Trade — The big picture

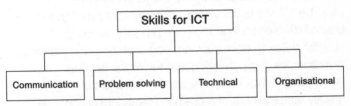

Most ICT jobs need a mixture of all four skill areas. Some jobs stress one or two skills more than the others. Some jobs need extra skills that are not shown here.

Now let's get down to the nitty-gritty.

If you say 'ICT job' to most people the first thing they will probably picture is someone in front of a screen or fixing a motherboard, so it's perhaps surprising that the thing the careers websites mention most is **good communication skills** – the ability to express yourself clearly to other people in words or writing. This doesn't just apply to jobs where you will be selling or training, but across the board. It makes sense when you think that most computer projects are too large for one person. People have to provide accurate information to one another about what is happening all the time or costly errors can result.

Many jobs involve a mixture of **teamwork skills** and the ability to **work alone**. So for example a software developer or web designer may well spend much of their time in front of a screen, but also work with other team members to see how different parts of the job affect each other, or attend meetings to discuss changing customer requirements. It helps if you are equally happy doing both.

The fast pace of business in ICT means that most staff must be able to **work under pressure** and **meet deadlines**. Games developers need to have products ready by specific dates to coincide with manufacturing and planned launch dates. If a network goes down, the network manager and systems administrator have to put it right quickly as the whole of a business may depend on it. In almost any ICT job, you need to be **flexible** enough to change your work plans to meet new requirements, and **adapt quickly to change** in the longer term, as fast-moving technology requires you to learn new skills all the time.

> 'I have learned that continuing professional development is essential – without it my IT skillset could be outdated in as little as three years.'
>
> *Andrew Margetts, Business Systems Analyst and Project Manager, Essex*

It also helps if you **don't panic** easily as the other side of this coin is that working in ICT generally requires **thoroughness** and **attention to detail**. Data entry clerks have to work carefully or someone may get a bill for £1,099 rather than £10.99! Even one character of incorrect code is often enough to make a programme crash. Whether working on hardware or software, you have to be **methodical** and **analytical** and have good **problem solving skills.** Much of the work is about using your brain to work out what is happening before you can find the best solution.

So perhaps it goes without saying that it helps if you are **well organised**, especially as in many jobs you will need to **juggle priorities** to work on many things at the same time. A support technician is likely to be working on several different problems for different people – who often all want an answer straight away!

If you look at some adverts for computer jobs, you will frequently see a bewildering list of acronyms and names of programmes, systems and qualifications the applicants ideally need to offer (e.g.: CISCO, asp.net, PERL, MSE, XML, UNIX etc). This highlights the fact that most jobs require **a variety of technical skills**. One or two won't do, so it's probably better to have some knowledge of as many as you can, rather than a detailed knowledge of just a few.

Notice that the focus in these adverts is almost entirely on skills. Few job adverts will say that you absolutely have to have taken certain subjects at school. It may help if you've done computing or ICT post-16, but not studying them does not stop you going for a career in ICT. The only real exception is the more technical jobs like computer engineer or telecommunications engineer, where a background in **maths** and **physics/science** is going to be essential. In many cases there are no laid down minimum educational qualifications.

DID YOU KNOW?

Several large computer firms including Microsoft have been employing people with autism to help with software testing. Those diagnosed with Autistic Spectrum Disorders tend to have better than average attention to detail and many have a photographic memory.

Source: Computer Weekly

On the other hand, you probably noticed from 'What's the story?' that most people in the industry have a **degree**, and there are some jobs that are very difficult to get into without one. Even then there are many employers recruiting graduates who have degrees in biology or German, not just ICT! Often employers are more interested in the 'brain power' we spoke of earlier than the actual degree subject. Deciding whether or not to do a degree is obviously important for you, but there are other ways in.

Unlike many jobs, there are no major health requirements either. If you want to work with hardware or telecommunications equipment, you will need to have **good eyesight and colour vision** but many jobs are mainly office based and often extremely suitable for people who have physical or other disabilities.

For some jobs, there are some more specific skills that you may need. If you want to be a games or multimedia designer, and to some extent a web designer or developer, you need to have some **artistic or creative ability** as well as a familiarity with specialised graphics packages. Web editors and technical authors need **writing skills** as much as ICT skills and you need to be **good with your hands** for jobs working with hardware. For the more customer-facing jobs, you may be using **sales** or **negotiation skills**. Several jobs, such as Sales, ICT trainer, and web designer amongst others can involve **presentation skills**.

All ICT staff will be trained on **health and safety issues** but some jobs require an understanding of **some aspects of law** – for example a database manager, who needs to have a thorough knowledge of the Data Protection and Freedom of Information Acts.

Lastly, some jobs need people who have a good **understanding of business** and this is particularly true of business analysts, project managers and many of the promoted posts the industry has to offer. The importance of business awareness is increasing as ICT becomes central to the way many companies operate.

SO WHERE DO YOU FIT IN?

Now you've found out about all the skills involved in ICT, how do you fit in?

Try this fun quiz to find out. Answer yes or no to the following questions, thinking honestly, how you normally react at home or at school:

1. Are you good at providing accurate information to other people?
2. Can you work equally well in a team or alone?
3. Can you stay cool when under pressure to get things done?
4. Are you good at solving problems (e.g. puzzles, crosswords, sudoku)?
5. Do you like to find out why things won't work?
6. Can you spot important details in a mass of information?
7. Would you describe yourself as 'well organised'?

If the answer to all or most of these questions is yes, then maybe you've got what it takes to succeed in ICT. If you are a real people person, an arty type or like fixing things with your hands, these points may give you some extra tips about the ICT jobs you might prefer.

Last of all, don't worry if you are not sure what skills you have yet. The saying goes that 'life's for learning.' Just keep thinking about the experiences it brings you and what they tell you about yourself.

DEGREE OR NOT

Of all employees in the ICT industry, 55% have a university degree – and the proportion is rising steadily. If you look at the ICT career articles on websites like jobs4u or the online Careers Advice Service, you will often see comments indicating that for a particular job, most/many entrants have a degree or comments like: 'A degree may be helpful.'

This is not the whole story as many of these articles will often add '… although there are no formal minimum entry requirements' but it is true that employers are looking for higher qualifications because more complex technology is demanding ever higher skill levels.

So should you aim to go to university, even if you do not really want to stay on in education? Here are some things for you to think about before you decide:

- Although the need for a degree is increasing, it does vary greatly between different ICT jobs. Nowadays, it is very unlikely that you will become a business analyst or systems developer without a degree, but there are still occupations like helpdesk operator or ICT technician that you can enter with lower qualifications.
- Often it is not the degree but the **skills** that it gives you that really count. Doing an ICT degree may not be the only way to get them. Many ICT graduates do not go into

THE VALUE OF A DEGREE

'Degrees are worth considering as lots of companies offer graduate programmes and we at Microsoft UK recruit 30 graduates a year. But you don't have to study Computer Science to work in ICT. Subjects that encourage logical thinking like sciences, engineering and maths are a good starting point too. There are also a lot of non-technical opportunities in ICT companies, such as human resources, finance and sales, which people often forget.'

Stephen Uden, Head of Skills and
Economic Affairs, Microsoft UK

ICT jobs while some employers prefer to recruit graduates who have **not** studied ICT.

- You may not want to do a degree after you leave school, but you may change your mind later on – perhaps when you see more clearly where it can lead you.
- If you don't like sitting behind a desk, training routes like the Modern Apprenticeship and the new IT Diploma provide practical ways of learning relevant skills.
- There are plenty of examples of computer whizzkids who have become very successful – for example as games designers or by starting an internet business. Many of them taught themselves, using 'how to' books and websites or short courses offered by software and hardware suppliers.
- If you run your own business, **you** and no one else will decide what qualifications you will need!

JARGON

Motherboard – The main circuit board in a computer, with slots to make different parts of the computer work, like the main hard drive, DVD mouse and keyboards.

Presentation Skills (as opposed to personal presentation!) – Presenting information to a group of people, often to help them make some sort of decision. Nowadays usually involves a computer programme such as PowerPoint.

7

Making your mind up

So here's what we have so far: a massive, growing industry with lots of jobs and a long list of different skill needs and the whole lot are changing all the time. Perhaps you're already starting to think that you have the skills to 'cut it' in an ICT job and like the 'buzz' of excitement that goes with it, or maybe it all seems a bit foreign and you still feel undecided.

Either way, don't stop reading yet. In the next part of this book we turn things around a bit. Instead of thinking all about the industry and its needs, let's think about you and your needs by looking at several frequently asked questions to help you decide on the one big question that's behind them all: 'what's in it for me?'

HOW CAN I GET MY FIRST JOB IN ICT?

If you're looking for an IT Apprenticeship, your Connexions personal adviser or careers adviser can give you details of what's available in your area. Learning and skills councils often produce lists of employers and training companies offering apprenticeships in your area.

'What are the jobs?' earlier in this book will give you an idea of the jobs that are likely to occur more generally, highlighting the ones that require lower levels of experience. If you reckon you will need more training before you apply, take a look at the chapter on 'Training for tomorrow'.

As well as the local and national papers, jobs get advertised in magazines like *Computing* and *Computer Weekly*. Many job websites and employment agencies carry ICT job ads and a site like www.agencycentral.co.uk will help you find them quickly. Then it's a case of picking the ones you like best.

Don't be too disappointed if jobs advertised in such places are asking for a lot of previous experience, even for 'junior' or 'trainee' vacancies. If you think you've more or less got the key skills they want, give it a try. The worst thing they can say is 'no' and many companies, especially smaller ones, will like a direct approach even if they haven't advertised a vacancy.

Application by CV and covering letter is standard, so it's important to make sure that you have a good CV and know how to use it. Some of the jobsites and agencies will give help on how to write a CV, or you can use a site like www.alec.co.uk/cvtips, or ask at your local Connexions or careers office for free advice.

ICT people like ICT solutions so another very common way of applying for ICT jobs is the online application form. Completing online applications is a bit different from paper forms, so watch out for the 'tricks of the trade' (below) as well as remembering all the other tips you have had about filling in application forms:

Online applications – tricks of the trade
- You will be given a username and password – so don't forget them!
- Not all online forms are automatically saved – make sure you keep a copy of what you have written just in case the system crashes.

- If the system allows, and not all of them do, make sure that you look over the whole form before you start writing.
- It may be possible to write your answers in a word processor and paste them into the form.
- Check if you are expected to complete the whole form in one session. If so, make sure you have all the information you need before you start.
- Keep your language as formal as you would for an offline form. Just because you're online doesn't mean you can lapse into 'text speak'.
- Some online forms include a brief selection test or introductory questions. If you don't get enough right answers, it could be 'game over' before you even start!

DID YOU KNOW?

As many as 65% of ICT employers offer their graduate trainees the chance to work abroad within the first two years.
Source: TARGET Graduate Trends Survey 2005/6: IT sector

WILL I NEED TO MOVE AWAY FROM HOME?

Not necessarily. As we saw from 'What's the story?', there are more ICT jobs in London and the South East, so it does help if you're prepared to live there. But there are jobs across the UK and not just in specialist ICT companies like IBM, Microsoft and EDS. Large numbers of ICT jobs come up in so-called ICT user companies. Almost all companies now use ICT in some way, so the list is almost endless but some obvious examples might be:

- Finance companies including banks, building societies and insurance companies
- Public sector bodies, such as the NHS, the Civil Service and local authorities

- Retail and wholesale companies
- Firms in the transport industry, airlines and airports
- Gas, water and electricity companies
- Schools, colleges and universities.

ICT is now a global industry so there are plenty of opportunities for you if moving away from home seems like a good idea, including working abroad in an increasing number of other countries. How would you fancy working in the USA, or Australia, or even China?

WILL I BE STUCK IN AN OFFICE?

Most ICT staff will be based in a nice, comfortable office environment but how much time they actually spend in it will vary a good deal from job to job. Helpdesk operators are likely to spend long periods at their workstation answering calls over a headset, whereas engineers, web developers and others will often need to be out of the office a good deal, perhaps staying away from home for several days or more. You might get the chance to work from home as well.

Very few ICT jobs would give you the chance to work outdoors much but, because of the need to work as a team, most ICT staff will spend a good deal of their time away from their desk in meetings. Many would also tell you the challenges of their job are more than enough to stop them getting bored.

WHAT'S THE PAY LIKE?

Generally well above the UK average. You won't feel like you've won the lottery if you start an apprenticeship but learning and skills councils tend to insist that apprentices are paid a minimum rate (£80 a week in 2008) and some companies will pay a lot more.

Figures from the regular salary survey of ICT workers make happy reading if earning good money is important to you. They suggest average salaries for qualified staff ranging from £23,041 for PC support to £104,707 for an IT director (March 2008). If you've looked at some of the job adverts, you will also have seen that a lot of jobs also pay bonuses and some offer other fringe benefits like a company car, pension or healthcare scheme.

DID YOU KNOW?

ICT professionals in the UK are among the best paid in the world. Salaries here rank as the fourth best overall, after Switzerland, Denmark and Belgium.
Source: Computing

These figures do have to be read with a bit of caution. Not all employers respond to surveys anyway and salaries in the South and in big cities are usually higher than elsewhere. Some of the better-paid staff are also on short-term contracts. So bear this in mind before you apply for a bank loan on the strength of your future earnings!

WHAT ABOUT THE HOURS?
These can vary quite a bit from job to job.

Many office-based ICT staff will work the typical "9 a.m. to 5 p.m." day most of the time. On the other hand 'on call' duties or tight deadlines mean that people sometimes have to work shifts or long hours including evenings and weekends. Database or systems administrators may have to be available at very short notice to solve a crisis just as designers and developers may need to put in extra hours to make sure products are ready on time.

There are some signs of a 'long hours' culture in the industry, particularly in the games sector. At the same time, prospects (www.prospects.ac.uk) say graduates working in the industry feel they have a good 'work–life balance'.

Holiday entitlements are also likely to vary from employer to employer.

HOW ABOUT MY LONG-TERM CAREER PROSPECTS?

Basically career prospects are excellent. Obviously in an industry like ICT that never stands still, the last thing you can expect is an automatic promotion ladder or a guaranteed job for life. We don't even know if the same job titles will be around in 20 years! But you can see from 'What's the story?' that the industry has a huge and growing demand for managers and specialists and for many people already, an ICT career has lead to top boardroom positions.

HOW DO I BUILD A SUCCESSFUL ICT CAREER?

The vital thing to realise is that your promotion prospects will depend totally on the **skills you can offer**, which is why it is so important that you always keep on learning new ones throughout your ICT career. It's even better if you use your **networking contacts** so that you can ring a few people to ask who's hiring and suss out the market. And it's better still if you can **spot the next big opportunity** that's coming along, giving you a head start when deciding what new skills you need to pick up. Unless you want work in multimedia design or computer engineering, these three success factors will probably be more important than learning any specific subject while you're still at school.

For example any software engineer who spotted the rise of the social networking sites like Facebook, Bebo and the likes a few years ago could have done their career a lot of good by learning about AJAX, the scripting language these applications are written in. Currently, the talk is all about acquiring general business skills and making better use of data analysis to shape business processes if that gives you any clues…

THE INTERVIEW

'At interview employers will look at how good you are at dealing with people as much as your technical skills. You also need to come over as positive, focussed, hard working and interested in developing your skills.'

Louise Mitchell, Programme Manager, Zenos, Banbury

WILL WORKING IN ICT CHANGE MY LIFESTYLE?

Probably any job will change you in some way and you may work long hours, but the 'price' of going into ICT is generally not high. Most jobs do not involve much physical work and risks are usually restricted to issues like electrical safety, use of VDUs, eyesight and repetitive strain injury.

Unless you are in the Armed Forces, it is unlikely that you will be asked to wear a uniform and you probably will not be expected to behave in a particular way outside work beyond making sure you respect the confidentiality of the information you are working with.

Many ICT people believe deeply that technology is there to help people and are prepared to 'go the extra mile' to make sure it does. One small problem you may find is explaining exactly what you do in your job at parties. Once you get going, most people won't have much clue what you're talking about!

8

WIIFM – what's in it for me?

Is this what you're looking for? Graduates working in the industry say the key advantages of their job are:

- Constant challenges and changes
- Enormous variety of work
- Relaxed work environment
- Culture of delivery and getting things done
- Part of a forever evolving and fast-moving industry
- Enthusiasm and intelligence of colleagues
- Good work–life balance.

Source: Graduate Prospects Sector Guide – Information Technology

Sample ICT Salaries		
Job title	**Average salary**	**Annual change 2006–2007**
PC support	£23,041	+5%
Programmer	£27,876	+11%
Systems administrator	£36,663	+3%
Database administrator	£40,024	+3%
Systems developer	£40,898	+4%
IT manager/consultant	£68,845	−1%

Source: Computer Weekly (March 2008)

JARGON

AJAX – Asynchronous Javascript And XML. A scripting language used by interactive social networking websites. A key feature is that it enables small parts of a webpage to be amended by the user without refreshing the whole page (but it also brings some security loopholes).

XML – Extensible Markup Language. A flexible system of common markup standards enabling data to be shared on the world wide web and elsewhere.

HTML – Hypertext Markup Language. The first language used for writing websites is basically a form of XML but uses pre-defined markup standards.

9

Success story 2

GARY – TECHNICAL DIRECTOR

From leaving school straight after his GCSEs, 35-year-old Gary is now Technical Director of System IT Ltd, a Microsoft Gold Partner certified company based just outside Carlisle. The company supplies, installs and maintains network servers, computers and multifunctional office equipment as well as provides IT training to businesses in the local area.

How did you first get interested in IT?
'At 16 I was really interested in electronics and did a GCSE in Technology. I was going to do a BTEC Electronics course at Carlisle College but for whatever reason decided to change to the BTEC National Diploma in Computing at the last minute. I did well on the course and my tutors advised me to go on for a degree. So I actually started a degree in computer science at the University of Staffordshire but unfortunately I got glandular fever at the beginning of the second year and had to take the rest of the year off to recover.

It's a very varied job and every day is different. I have to prioitise well, as there are lots of instant demands coming in.

'Back at home in Carlisle, I was offered a job as a Data Entry Clerk with some programming work at Cumbria County Council. By the end of the year I was doing more and more programming and decided not to go back to university.'

How did your IT career progress from there?
'I spent four years on the programming team at Cumbria CC but at that time computer staff in local authorities weren't well paid and so I went for a pretty junior IT engineering job with Criffel Systems in Dumfries, fixing PCs and printers.

'Criffel won a contract with NHS Scotland to update IT systems in doctors' surgeries. There was a big expansion and I ended up spending all my time replacing old-fashioned, green screen PCs with Windows-based systems and new network servers in every doctor's surgery across Glasgow and South West Scotland. By now I was getting well paid as a proficient engineer, but I was also driving 40,000 miles a year.

'Through doing a course with System Driver Training Ltd I got to know the directors of System Group. I had watched their company grow at a terrific rate, and after some discussion I was offered the chance of setting up a new company within System Group, System IT Ltd. System Driver Training was a company I already had my eye on but the new job meant a £10,000 drop in salary. After discussing it with friends and family, I decided to accept and became Director of System IT. The company started to grow and after two years of steady progress we brought a sales director on board to start the office machinery side of the business, and to grow the business to a new level.'

What does your job actually involve?

'I handle the sales of IT equipment for our customers, co-ordinate installation projects, and give advice to our engineers. I do the ordering and purchasing and deal with the suppliers. I'm very "hands-on" and still work on installations and respond to support calls. I'd hate to spend all my time in the office and am lucky enough to have a fellow director who covers most of the administration and staffing issues.

'It's a very varied job and every day is different. I have to prioritise well as there are lots of instant demands coming in. I also enjoy seeing the company grow. We now have several large local customers including solicitors, schools and Carlisle United Football Club.

'The downside is the pressure. I work 12-hour days, often including some time at weekends as we offer a 24/7 instant response service. It can be difficult to find time to relax. I also have to keep learning to keep my own technical skills up to date so that I'm one step ahead of the engineers – and some of them are very sharp!'

What is your ambition?

'I expect to spend at least the next 15 years growing the company.'

What advice would you give young people who are thinking of working in IT?

'Everyone needs IT. It's an exciting area to look at with lots of varied fields [see 'What are the jobs?']. You can also make good money and some contractors earn as much money in 30 weeks as employed staff do in a full year.

'The IT Apprenticeship programme is a good way forward for people who do not want to do a degree, but want to start earning and learning at the same time. When interviewing for new staff, I rate Microsoft technical qualifications and experience above a degree anyway because they are more practical.

'Qualifications like the BTEC National Diploma I did are nice and practical too but you have to find a job for yourself at the end of it whereas most apprentices are kept on by their employers.'

JARGON
Microsoft Gold Certified Partner – A trade recognition award. Gold Partners are formally assessed by Microsoft as achieving the highest level of technical competence and customer service in implementing their systems.

Source: Microsoft

10

Training for tomorrow

If you need to learn new skills throughout your career in ICT, it's obvious that training will pay a big part in building your future. Like the rest of the industry, the ICT training scene is moving on pretty quickly. From outside, it looks a bit like a big experimental greenhouse in which colourful new plants are growing all the time. Some of the plants will flourish, but others will probably die back and some will only grow in the correct soil. Again it's down to you to decide which are right for your own career garden and pick the ones that will help **you** grow.

SCHOOL QUALIFICATIONS

As you know, it is now compulsory to study some ICT and the GCSE double award in ICT and Applied ICT is an obvious way of learning as much about ICT at an early stage, as is the Standard Grade in Computing in Scotland.

At AS and A level it is possible to choose between the more general **ICT** and **Computing**, which is perhaps more for those who have already decided they are on their way to becoming ICT professionals. In Scotland, **Information Systems** and **Computing** are separate subjects offered from Intermediate to Advanced Higher.

New Information Technology (IT) Diploma (England)

September 2008 saw the introduction of a totally new type of qualification in schools for 14–19-year-olds – the vocational Diploma is available in a variety of subjects including IT. Diplomas will be available at Level 1 (equivalent to 4–5 GCSEs grades D–G), Level 2 (equivalent to 7 GCSEs grades A*–C) and Level 3 (equivalent to 3.5 A levels). A mixture of practical learning and theory, the IT Diploma will develop your ICT knowledge and skills as well more general skills to help your future learning and employment.

The Diploma will be taught by a number of training providers, not just schools. Employers will be involved, perhaps providing visits, work experience, mentoring, classes and projects or case studies. All Level 3 courses will include some form of work experience but it will still be possible to study them alongside traditional A levels.

The IT Diploma is being run in a number of areas from 2008 and a full list is available at www.e-skills.com

The IT diploma in Wolverhampton

One area where the IT Diploma is already available is Wolverhampton, where 14- and 15-year-old pupils will be able to spend two days a week studying for it alongside their core GCSE subjects, even if it is not being run in the school they attend.

Peter Hawthorne of Wolverhampton City Council's 14–19 Development Team stresses the importance of making sure that young people are well advised before they enter the programme:

'We run a number of CARD (Choose A Real Deal) taster activities which give students the opportunity to see what each Diploma is really like. It is an expectation that diploma students will have undertaken Diploma CARD taster activities before going on to the next stage, the Online Progression Preference process, which learners interested in Diplomas fill in online.

'This allows the suitability of the learner for the specific area of study to be tested. Appropriate guidance for all learners expressing an interest is a really important part of this process to ensure all learners are guided to curriculum choices that suit them best.'

Successful applicants will learn how to deliver successful projects, creating ICT solutions to meet the needs of business and developing their ability in a professional setting.

Employers likely to be involved in Wolverhampton include Carillion plc and Persimmon Homes. Nationally, big ICT companies like LogicaCMG, Microsoft, Vodafone and Cisco have all promised their support.

IT APPRENTICESHIPS

IT Apprenticeships are a vital stepping stone for young people who want to get into the industry as soon as possible and a lot of people have had their eyes opened to the jobs potential in the industry through completing one.

There are two main levels of apprenticeship. Level 2 Apprenticeships (Skillseekers in Scotland) are available to all leavers subject to specific exam requirements from employers, whereas Level 3 Advanced Apprenticeships (Skillseekers Modern Apprenticeships) are only available for those with A levels or equivalent qualifications. A third type, the Higher Apprenticeship, offers the chance for trainees in

England to 'fast track' onto a degree course by combining study for a Foundation Degree (see below) with work experience.

The main kinds of IT Apprenticeship are:

The **IT User (iTQ)** Apprenticeship aims to give you the right user skills for the kind of work you are doing using a 'pick and mix' approach. It could cover anything from web design to working as a bank clerk. A popular choice with both young people and employers, this is the kind of apprenticeship Rob did (see Case Study 1).

An apprenticeship in **ICT Services and Development** is for more technical roles while **Communication Technologies (Telecoms)** covers the telecommunications side of the industry. There are big overlaps between telephone and computer technology, meaning that many apprenticeships cover both areas. An example is the Advanced Apprenticeship offered by mobile phone company Orange, which also requires study for a college qualification.

Contact Centre apprenticeships focus on dealing with customers and could cover helpdesk operations in user companies such as travel firms, sales or the emergency services as well as ICT companies.

'If you are looking to kick-start your IT career an apprenticeship is possibly the best springboard as you will "earn as you learn".'

Rob Langer, Media Technician,
Staffordshire County Council

NVQs and all that

Traditionally, the piece of paper you got at the end of the apprenticeship was an **NVQ** certificate at Level 2 or 3. This is still the case in Scottish education, where the best guide to vocational qualifications is the **SVQ Update**. You can download the latest copy from the Scottish Qualifications Authority website at www.sqa.org.uk.

Otherwise, NVQs are being gradually replaced by **VRQ**s (Vocationally Related Qualification) but you may encounter both terms in connection with ICT training for some time to come. Whereas NVQs focus on your competence to do tasks in the workplace, VRQs are more about assessing the knowledge and understanding you have acquired.

NVQs and VRQs come in all sorts of shapes and sizes and are mainly offered by Edexcel, City & Guilds and OCR. For a full list, check out the National Database of Accredited Qualifications at www.accreditedqualifications.org.uk. Don't worry too much about the bewildering number available as your training provider will help you decide which ones you actually do.

> **DID YOU KNOW?**
>
> The careers section of the e-skills website www.e-skills.com contains more information on current ICT qualifications, the IT Skills Passport and a list of training providers for Apprenticeships across the UK.

Integrated into the ITQ, the **IT Skills Passport** (PC Passport in Scotland) from e-skills UK is a lifelong record of achievement for IT user skills, which helps you to assess your current IT skills and identify gaps so you can work out what sort of training you will need in future and show employers what you can do already.

How do I get a VRQ?

They are delivered through a taught programme and have a work placement running alongside – this gives you an opportunity to use what you are learning and develop your skills as you work through the programme. To achieve a VRQ you are required to complete a series of set tasks and exercises that will help demonstrate your knowledge and understanding, each of which will ask you to reflect on and give real examples from your practice – giving you the opportunity to show that you can use your knowledge and understanding in your work. *Source:* National Youth Agency

FURTHER EDUCATION

Further education courses like the one Gary studied (see Case Study 2) are still a good way of getting a useful qualification in ICT skills. Key options are the full-time BTEC National Diploma courses in System Support, Software Development, Networking, and IT and Business, or the National Certificate modules in IT and Computing subjects in Scotland. Part-time options are also available for those with work or family commitments.

Not everyone likes the discipline at school and some people come on leaps and bounds when they get the chance to study work-related subjects at a college where they feel they are less restricted. These days it's quite possible to combine both school and further education courses anyway and Educational Maintenance Allowances mean that there is no need to lose out financially if you continue your education.

For more information on college courses, ask at your local college or see a copy of the *Directory of Vocational and*

Further Education at your local Connexions office, or library. For Scottish courses, you can also search at www.learndirectscotland.com

HIGHER EDUCATION

Depending on your exact qualifications, there are now an extremely wide variety of HND and degree courses in ICT subjects across the UK. Nowadays, you can do degrees in quite specialised aspects of ICT, including computer games technology, computer aided drafting, forensic computing, artificial intelligence, e-business to mention just a few. You can do a course search at www.ucas.com to find out what degree courses will suit you.

> ### DID YOU KNOW?
>
> The University of Abertay has developed an MSc postgraduate course in Ethical Hacking and Countermeasures
> Source: University of Abertay website

FOUNDATION DEGREES (ENGLAND)

There are now over 300 Foundation degrees in computing subjects in England, including specialisms such as networking and games development.

Foundation degrees are a work-related higher education qualification designed to meet employer skills shortages and many students are studying the Foundation degree part time alongside their ICT job. They are usually delivered at FE colleges but allow entry to the second or third year of a degree at university. Experience counts for more than qualifications but some courses do accept students with one A level. The best place to find details on Foundation degrees is www.fdf.ac.uk.

Main routes into ICT employment

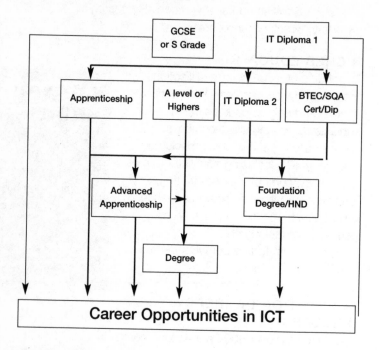

SUMMARY

The above diagram shows the main routes that you can follow from school through to employment in the ICT industry. The broad lines show the most obvious routes.

BECOMING A PROFESSIONAL

As you can guess already, there is no firm dividing line between courses and qualifications you do before and after getting your first ICT job, as with many other jobs.

ICT is unlike some professions in that you do not have to join a particular professional body as you would if you wanted to be a legal executive or accounting technician. Some ICT professionals decide to do the examinations offered by the British Computer Society and some decide to join particular professional organisations to learn more about their own ICT specialism. Examples are the Institution of Analysts and Programmers, the Help Desk Institute or the UK Web Design Association.

VENDOR QUALIFICATIONS

Often more important in the ICT world are the so-called 'vendor' qualifications issued by the big hardware and software companies and especially Microsoft. Basically vendor qualifications train you in the use of particular products sold by the company and you may be required to have passed the qualification before you can use the product.

Key qualifications offered by Microsoft are Microsoft Certified Professional (MCP), Microsoft Certified Systems Engineer (MCSE), Microsoft Certified Systems Developer (MCSD) and Microsoft Certified Trainer (MCT). Important qualifications from other companies are Cisco Certified Instructor (CCI), Cisco Certified Network Engineer (CCNE), Novell Network Authorised Engineer (NNAE), the ORACLE Database Administrator series and the COMPtia A+ (for PC Support) and Network + certificates.

> **DID YOU KNOW?**
>
> Many recruiters have noted that men will put themselves forward for a job for which they are 50% equipped. Women will apply only if they feel 100% qualified.
> Source: Getting Women on Board. The Business Case for Diversity, Joanne Welch, Verdus, March 2005

Colleges and apprenticeship providers now also offer vendor qualifications as well as companies like Computeach, who train people looking to change to an ICT career. You can pay to do them yourself but they are very expensive and you should make sure they are not more detailed than you really need.

TEACH YOURSELF COMPUTING!

Perhaps a simpler and cheaper route if you are just starting to learn the technical skills is to teach yourself using one of the many websites or self-help books that will take you step by step through some of the key ICT skills. Sites like www.webmonkey.com and 'how to' books in your local library or bookshop can provide you with an excellent introduction to all kinds of important ICT skills from basic web design to networking and project management.

'After I'd learned a few IT skills, I found I could pick up product manuals and work out how to do most things after wading through them.'

Andrew Margetts, Business Systems Analyst and Project Manager, Essex

---

WHICH WAY WILL YOU GO?

Tick one box to show which way you think you will choose to find your route into an ICT job?

1. Stay on at school and go straight to university? ☐

2. Leave school and go to college? ☐

3. Go for an IT apprenticeship? ☐

4. Get a different job and work my way into ICT later? ☐

5. Start my own business right away? ☐

It doesn't matter which one you picked, but do try to keep an open mind. Your plans could well change over time.

ICT WILL NEED MORE WOMEN

The alarming drop in women working in ICT occupations – down to 14 or 16% in 2007 according to some estimates – is starting to concern companies who can see they may struggle to find enough staff in the future.

Many companies are starting to look at how they can become more 'woman-friendly' but research also suggests that girls may be turned off ICT at school because they see it as too 'nerdy' or confuse it with more traditional jobs like typist or office clerk.

There is now a lot of help if you are a girl who is interested in an ICT career. At school, you can often join in with Computer Clubs for Girls (CC4G – www.cc4g.net) to enjoy fun ICT projects about music, fashion, celebrities and crime-scene investigations. Along with the boys, you can also experience projects like Go4IT for Year 9 pupils and Experience IT Chances in Years 10 and 11. More details are available at www.e-skills.com.

DID YOU KNOW?

The Office for National Statistics' Survey of Hours and Earnings for 2006 indicates that 18% of IT jobs are held by women. In 1997 it was 27%.

ARE GIRLS MISSING OUT?

'Although I notice that not many girls apply for our Computing courses, they seem to be at an advantage when it comes to applying for jobs afterwards. I think many ICT employers are convinced that recruiting women will broaden the range of skills their company will have available in the future.'

Glenn Affleck, Programme Leader,
BSc Computing, University of West of Scotland

You needn't feel like the only girl in the computer room as there is no shortage of help for young women wanting to make a career in the industry. The British Computer Society Women's Group at www.womenintechnology.co.uk offers networking events, news and events listings and careers advice for women working in technology. Intellect UK's *Women in IT Forum* www.intellectuk.org gives a good idea of the companies who want to recruit more women and www.equalitec.org.uk works with women coming back to work after a career break.

Meanwhile, *Where Women Want to Work* www.www2wk. com is a website of jobs including many ICT opportunities and *Wise* www.wise-women.org supports women who want to become web designers, developers, and programmers.

Many women have already made it to the top in ICT, running their own businesses or occupying boardroom positions in major national companies. There's no reason why you cannot do the same if you are determined to succeed.

MORE WOMEN AT MICROSOFT

'At Microsoft UK, 27% of our staff overall, and 50% of our graduate trainees, are women. We haven't targeted women specifically but all our employees can work flexible hours, including working from home, which enables women to benefit. We also offer crèche facilities and "bump clubs" where pregnant women can access advice and support.

'Schools in the Thames Valley area can take part in our "Digigirlz" initiative. We invite the girls into the company to speak to the women who work for us and show them the technology we are using, especially the new online businesses to show them that they will not necessarily be working as a boffin in a white coat.'

Stephen Uden, Head of Skills and Economic Affairs, Microsoft UK

JARGON

NVQ – National Vocational Qualification.
SVQ – Scottish Vocational Qualification, the Scottish equivalent of NVQ.
VRQ – Vocationally Related Qualification, gradually taking over from NVQ.
ITQ – Information Technology Qualification.
UCAS – Universities Central Admissions Service, which administer applications to degree courses in the UK.

11

The last word

By now you should have a good idea of what the ICT industry looks like at the moment. You've probably already picked up that it's going to change pretty fast in the future. Perhaps you could compare a career in ICT to going for your first big fairground ride – it's certainly fast and exciting, possibly a bit scary and you never quite know what's round the next corner. On the other hand there will be times when you need to be persistent and determined, like a showperson who keeps practising until they've mastered a new trick.

To help you get the best from the ICT fair, here are **ten top tips** to bear in mind if you decide to make it your career:

1. Build on your ICT skills: focus on areas considered to be in demand in the industry.
2. Use the skills profiling tools available and work on your weaker skill areas.
3. Promote yourself: don't undersell any ICT-related knowledge.
4. The fast pace of change means you must be prepared to change too.
5. Get as much work experience as possible.
6. Teach yourself computing – by using sites like webmonkey and 'how to' books.
7. Make up a CV if you haven't already and find out about online applications.

8. Use the recommended websites and publications to check out course and job opportunities.
9. Don't be scared to apply for jobs you are not totally qualified for – especially if you are female.
10. Consider the possibility of working freelance or running your own business one day.

If you have made it this far through the book then you should know if ICT really is the career for you. But, before contacting the professional bodies listed in the next chapter, here's a final, fun checklist to make sure you're making the right decision.

THE LAST WORD

✔ TICK TRUE OR FAL

I HAVE **MOST** OF THE SKILLS REQUIRED IN ICT (SEE QUIZ AT THE END OF 'TOOLS OF THE TRADE').

☐ TRU
☐ FAL5

I AM PREPARED TO KEEP LEARNING NEW SKILLS EVERY YEAR.

☐ TRU
☐ FAL5

THE IDEA OF CONSTANT CHANGE EXCITES ME.

☐ TRU
☐ FAL5

WORKING LONG HOURS IS FINE BY ME IF THE PAY IS RIGHT.

☐ TRU
☐ FAL5

I WOULD BE INTERESTED IN FINDING OUT MORE ABOUT HOW DIFFERENT BUSINESSES WORK.

☐ TRU
☐ FAL5

I BELIEVE THAT ICT IS THERE TO MAKE PEOPLE'S LIVES EASIER.

☐ TRU
☐ FAL5

If you answered 'TRUE' to all these questions then CONGRATULATIONS! YOU'VE CHOSEN THE RIGHT CARE If you answered 'FALSE' to any of these questions then this may not b the industry for you. However, there are still some options open to you. example, you could work in an office, a shop or a hotel and still have pl of chances to use a computer.

12

Further information

PROFESSIONAL BODIES AND ORGANISATIONS

The following all have information about careers or training in ICT. Some of them specialise in particular aspects, e.g. The Games Industry or in-company training:

British Computer Society (BCS), 1st Floor, Block D, North Star House, North Star Avenue, Swindon SN2 1FA. 01793 417417. Website: www.bcs.org

British Interactive Media Association (BIMA), Briarlea House, Southend Road, Billericay CM11 2PR. 01277 658107. Website: www.bima.co.uk

The Entertainment and Leisure Software Publishers Association (ELSPA), 167 Wardour Street, London W1F 8WL. 020 7534 0580. Website: www.elspa.com

e-skills UK, 1 Castle Lane, London SW1E 6DR. 020 7963 8920. Website: www.e-skills.com ***

Help Desk Institute (HDI), 21 High Street, Green Street Green, Orpington, Kent BR6 6BG. 01689 889100. Website: www.hdi-europe.com

---

*** denotes three star resource. These are **'must reads'** if you are serious about an ICT career!

Institute of IT Training, Westwood House, Westwood Business Park, Coventry, CV4 8HS. 0845 006 8858. Website: www.iitt.org.uk

Institute for the Management of Information Systems (IMIS), 5 Kingfisher House, New Mill Road, Orpington, Kent BR5 3QG. 0700 002 3456. Website: www.imis.org.uk

The Institution of Analysts and Programmers, Charles House, 36 Culmington Road, London W13 9NH. 020 8567 2118. Website: www.iap.org.uk

The National Computing Centre (NCC), Oxford House, Oxford Road, Manchester M1 7ED. 0161 228 6333. Website: www.ncc.co.uk

Skillset, Focus Point, 21 Caledonian Road, London N1 9GB. Tel: 020 7713 9800. Website: www.skillset.org

TIGA, Brighton Business Centre, 95 Ditchling Road, Brighton BN1 4ST. 0845 094 1095. Website: www.tiga.org

USEFUL WEBSITES
www.accreditedqualifications.org.uk
National Database of Accredited Qualifications

www.agencycentral.co.uk
Agency Central – directory of recruitment websites. Pick out the ones dealing with ICT vacancies

www.alec.co.uk/cvtips
Alec – plenty of advice and examples on writing CVs and covering letters

www.blitzgames.com/gameon
Blitz Games – careers in the games industry

Become instantly more attractive

To employers and further education providers

Whether you want to be an architect (Construction and the Built Environment Diploma); a graphic designer (Creative and Media Diploma); an automotive engineer (Engineering Diploma); or a games programmer (IT Diploma), we've got a Diploma to suit you. By taking our Diplomas you'll develop essential skills and gain insight into a number of industries. Visit our website to see the 17 different Diplomas that will be available to you. www.diplomainfo.org.uk.

www.bcs.org.uk/bcswomen
British Computer Society Women's Group

www.careers-scotland.org.uk
Careers Scotland

www.careersserviceni.com
Careers Service Northern Ireland

www.careerswales.com
Careers Wales

www.ciwcertified.com
Certified Internet Webmaster qualification

www.cc4g.net
Computer Clubs for Girls

www.cisco.com/web/UK
Cisco UK – vendor qualifications

www.comptia.org
Computer Technology Industry Association (CompTIA) –
vendor qualifications

www.connexions-direct.com/jobs4u ***
Jobs4u – careers information database covering most of the
jobs in plenty of detail

www.equalitec.com
Equalitec – for women in ICT and Electronics

www.fdf.ac.uk
Foundation Degree Forward – includes search for
Foundation Degree courses

www.igda.org
The International Game Developers Association

www.intellectuk.org/women
Intellect's Women in IT forum

www.careersadvice.direct.gov.uk/helpwithyourcareer
Learndirect England – including careers information articles

www.learndirectscotland.com
Learndirect Scotland – search for courses in Scotland

www.mentorset.org.uk
MentorSET – mentoring scheme for women in science,
Engineering and Technology

www.microsoft.com/uk
Microsoft UK

www.novell.com/training
Novell Training

www.prospects.ac.uk
Prospects – graduate level careers opportunities

www.sqa.org.uk
Scottish Qualifications Authority – including SVQ update

www.ucas.com
UCAS – university courses and applications

www.webgenies.co.uk
Web Genies – information and web development for young
people and beginners

www.webmonkey.com
Webmonkey – loads of useful tutorials on computing skills

www.www2wk.com
Where Women Want to Work – jobsite

www.wise-women.org
Women into Science and Engineering

www.womenintechnology.co.uk
Women in Technology

BOOKS, MAGAZINES, LEAFLETS etc

Careers in IT (leaflets from the British Computer Society)

Careers in Information Systems (Institute for the Management of Information Systems)

Computer Weekly (www.computerweekly.com)

Computing (www.computing.co.uk)

Develop (www.developmag.com)

Directory of Vocational and Further Education (Pearson Education)

Edge (www.edge-online.co.uk)

Inside Careers Guide to Information Technology (Inside Careers Guides – Genevieve Dutton)

Make Your Choice (Institution of Engineering and Technology)

MCV (www.mcvuk.com)

PC Pro (www.pcpro.co.uk)

Revolution (www.brandrepublic.com/revolution)

Workingames (www.workingames.co.uk)

Working in computers & IT *** (Connexions)